ELECTRONICS FOR BEGINNERS

Unveiling the Ultimate Guide to
the World of Electronics for
Absolute Beginners

CHARLES LIGHT

Table of Contents

CHAPTER ONE 3

    Introduction to Electronics 3

    Properties and applications of resistors, capacitors, and inductors...............................5

    Introduction to Integrated Circuits 6

    Fundamental Concepts in Electronics............8

    Ohm's Law and Basic Circuit Analysis 9

    Series and Parallel Circuits 10

CHAPTER TWO 11

    Introduction to breadboarding and prototyping 11

    Types and Functions of various ICs 13

    Introduction to microcontrollers 14

    Fundamentals of Arduino Programming......15

    Projects with Advanced Components and Circuits 16

    Binary System and Number Representation 17

CHAPTER THREE 19

    An introduction to flip-flops and sequential logic..19

    Basic Concepts in Digital Signal Processing..20

- Building a simple LED circuit 21
- Designing a Temperature Controlled Fan 23
- Building a digital thermometer 24
- Building a Basic Sound Synthesizer 25

CHAPTER FOUR 26
- Common Mistakes in Electronic Projects.....26
- Troubleshooting for Electronic Circuits 28
- Resources for more learning and project ideas...29

THE END 31

CHAPTER ONE

Introduction to Electronics

Electronics is the branch of physics and technology that studies electron emission, behavior, and effects in vacuum, gas, and semiconductors. It entails the investigation of electronic components, circuits, devices and systems.

Electronic components:

Resistors are passive two-terminal electrical components that block the flow of electricity. They are frequently used to regulate the amount of current in a circuit or to split voltage. They're measured in ohms.

Capacitors are passive two-terminal electrical components that store electrical energy within an electric field. They are frequently used to smooth out voltage swings in power supply or to block DC current while allowing AC current to flow.

Diodes are semiconductor devices that allow current to flow in a single direction alone. They are employed in rectifier circuits to convert alternating current (AC) to direct current (DC), as well as in signal modulation and voltage regulation.

Transistors are semiconductor devices capable of amplifying or switching electronic signals and electrical power. They are key components of modern electrical gadgets like computers and cellphones.

Inductors are passive two-terminal electrical components

that store energy in a magnetic field while current passes through them. They're often utilized in filters, oscillators, and power supply.

Properties and applications of resistors, capacitors, and inductors

Resistors are employed not just to control current, but also in voltage dividers, timing circuits, and heating elements.

Capacitors are employed not just to smooth voltage, but also in timing circuits, filters, and energy storage.

Filters, transformers, oscillators, and energy storage systems all require inductors.

Operating Principles and Applications of Diodes and Transistors

Diodes work by allowing current to flow in one direction (forward bias) while blocking it in the opposite direction (reverse bias). They are utilized for rectifiers, signal demodulation, and voltage regulation.

Transistors operate by directing the flow of current between two terminals (collector and emitter) via a third terminal (base). They are utilized in amplifiers, switches, oscillators, and digital circuits.

Introduction to Integrated Circuits

Integrated circuits (ICs) are miniature electronic circuits made up of semiconductor devices (such as transistors, diodes, and resistors) and passive components (such as capacitors and inductors) constructed on a single chip of semiconductor material.

ICs, which can contain thousands to billions of electronic components, are utilized in almost all electronic equipment, including computers, smartphones, medical devices, and automotive systems.

This summary should give you a general understanding of these key topics in electronics. Please let me know if you require more detailed explanations or examples.

Fundamental Concepts in Electronics

Voltage, Current, Resistance:

Voltage (V) represents the electrical potential difference between two locations. It is measured in volts (V) and represents the force that propels an electric current across a circuit.

Current (I) is the flow of electrical charge. It is measured in amperes (A) and denotes the rate at which charge flows across a circuit.

Resistance (R) is the opposition to the flow of electrical current. The material, length, and cross-sectional area of a conductor determine its resistance (measured in ohms).

Ohm's Law and Basic Circuit Analysis

Ohm's Law says that the current flowing through a conductor between two locations is exactly proportional to the voltage across those points and inversely proportional to their resistance.

Basic Circuit Analysis: Simple circuits are analyzed and solved using Ohm's Law and Kirchhoff's Laws (which deal with charge and energy conservation in electrical circuits).

Series and Parallel Circuits

A series circuit is one in which the components are linked end to end, leaving only one path for electricity to flow. Each component receives the same current, and the total resistance is calculated by adding the individual resistances together.

Parallel Circuit: A parallel circuit is one in which the components are linked across each other's terminals, creating numerous channels for current to flow. The voltage across each component is constant, and the total resistance is less than the lowest individual resistance.

CHAPTER TWO

Introduction to breadboarding and prototyping

Breadboarding is the practice of creating and testing electronic circuits without using solder. It includes quickly prototyping circuits on a breadboard, which is a reusable platform with interconnected sockets.

Prototyping is the process of producing a working model of a circuit or electronic device in order to evaluate its functionality and practicality before mass manufacturing.

Building Simple Circuits (LED Blinkers and Voltage Dividers):

LED blinker circuits typically include an LED, a resistor (to limit current), and a power source (such as a battery). By properly connecting the components, you can make the LED flicker on and off.

Voltage Dividers: A voltage divider circuit is made up of two resistors linked in series to a voltage source. It splits the voltage across the resistors based on their values and can be used for a variety of applications, including establishing a reference voltage and biasing a transistor.

Understanding these principles will help you grasp the fundamentals of electronics while also preparing you for more advanced topics. If you have any questions or require clarification, please ask!

Sensor and Actuator:

Sensors are devices that detect and respond to physical inputs. They convert physical quantities (e.g., temperature, light, pressure) into electrical signals.

Actuators are devices that use electrical signals to produce physical actions or movements. They are used to regulate or manipulate physical systems.

Types and Functions of various ICs

Amplifiers, filters, and voltage regulators are examples of analog integrated circuits that deal with continuous signals.

Digital integrated circuits use discrete levels of voltage to represent binary numbers. They comprise logic gates, flip-flops, and microcontrollers.

Mixed-signal integrated circuits (ICs) mix analog and digital circuitry and are commonly used in applications that require both forms of processing, such as data conversion and signal processing.

Introduction to microcontrollers

Microcontrollers are compact, self-contained computer systems that use a single integrated circuit. They include a CPU, memory (RAM or ROM), and I/O peripherals.

Microcontrollers are utilized in embedded systems to control many devices and systems, including domestic appliances, automobile systems, and industrial controls.

Fundamentals of Arduino Programming

Arduino is an open-source electronics platform with simple hardware and software. It consists of a microcontroller board and a development environment that allows you to write, compile, and upload code to the board.

Arduino programming is done with a simplified version of C/C++. You can create code to read sensors, control actuators, and respond to environmental signals.

Projects with Advanced Components and Circuits

Advanced projects may entail the use of sensors to collect data (e.g., temperature, light, motion) and actuators to respond to that data (for example, turning on a fan based on temperature).

You can also use ICs to build more complicated circuits for specific applications like as motor control, wireless communication, and data logging.

These subjects expand on the fundamentals of electronics and circuitry, enabling you to design increasingly complex projects and devices. If you're interested in specific projects or need more information on any of these topics, please ask!

Binary System and Number Representation

The binary system is a base-2 number system that represents numbers with two digits (0 and 1). A bit is the term used to refer to each digit in a binary integer.

Binary numbers are employed in digital electronics and computers because they are conveniently represented by two-state electronic switches (such as transistors).

Logic gates and Boolean algebra:

Logic gates are electronic circuits that apply logical operations (AND, OR, NOT, etc.) to one or more binary inputs to generate a single binary output.

Boolean algebra is a mathematical theory concerned with binary variables and operations. It is used for analyzing and designing digital circuits.

CHAPTER THREE

An introduction to flip-flops and sequential logic

Flip-flops are digital circuits that can hold a single bit of data. They are the building blocks of sequential logic circuits, in which the output is determined not only by the current input but also by the previous sequence of inputs.

Sequential logic is employed in applications where the system's behavior is determined by its previous inputs, such as counters, registers, and memory units.

Basic Concepts in Digital Signal Processing

Digital Signal Processing (DSP) is the manipulation of signals in the digital realm, which is commonly done with digital computers and specialized hardware.

DSP activities include filtering, encoding, and compression, and it is utilized in a variety of applications, including audio processing, picture processing, and communications.

Practical applications of digital electronics:

Computers, cellphones, digital cameras, and communication systems are all examples of applications that use digital electronics.

They are also employed in industrial automation, robotics, medical devices, and automobile systems, among other applications.

Understanding these ideas will provide you with a firm foundation in digital electronics, preparing you for more advanced courses in the field. If you have any specific questions or require extra clarifications, please ask!

Building a simple LED circuit

Components Required: LED, resistor, breadboard, jumper wires, and battery.

Circuit Design: Connect the longer leg of the LED (anode) to the resistor, then connect the other end of the resistor to the battery's positive wire. Connect the short leg of the LED (cathode) to the battery's negative terminal.

When the circuit is powered, the LED illuminates.

Create a Light-Sensitive Alarm:

Components Required: light-dependent resistor (LDR), resistor, buzzer, transistor, battery, breadboard, and jumper wires.

Circuit Design: Put an LDR in series with a resistor. Connect one end of the LDR resistor series to the transistor's base and the other to the battery's positive terminal. Connect the transistor's emitter to the battery's negative terminal. Connect the buzzer to the collector of the transistor and the positive terminal of the battery.

Function: The alarm will ring when the light falling on the LDR falls below a predetermined level.

Designing a Temperature Controlled Fan

Components Required: Temperature sensor (e.g., thermistor), transistor, fan, resistor, battery, breadboard, and jumper wires.

Circuit design: Put the thermistor in series with a resistor. Connect one end of the thermistor-resistor series to the transistor's base, and the other to the battery's positive terminal. Connect the transistor's emitter to the battery's negative terminal. Connect the fan to the collector of the transistor and the positive terminal of the battery.

The fan will switch on when the temperature detected by the thermistor surpasses a particular threshold.

Building a digital thermometer

Components Required: Temperature sensor (e.g., LM35), Arduino board, LCD display, breadboard, and jumper cables.

Circuit Design: Connect the LM35 to Arduino's analog input pin. Connect the LCD display to the Arduino's digital pins for data and control.

Function: The Arduino reads the temperature from the LM35 and shows it on the LCD as a digital thermometer.

Building a Basic Sound Synthesizer

Components Required: Arduino board, speaker, potentiometer, breadboard, and jumper wires.

Circuit design: Connect the potentiometer to the Arduino's analog input pin. Connect the speaker to the Arduino's digital output pins.

Function: The Arduino reads the potentiometer's analog input and generates the corresponding sound frequency, which is then output through the speaker.

These projects provide an excellent opportunity to learn and experiment with fundamental electronic components and circuits. They can be expanded and changed to meet your specific interests and learning objectives.

CHAPTER FOUR

Common Mistakes in Electronic Projects

Incorrect component orientation: Diodes and electrolytic capacitors must be connected in the right polarity.

Not referring to datasheets for components can result in inappropriate usage or failure.

Improper soldering: Cold joints, inadequate solder, or excess solder can result in faulty connections.

Incorrect wiring: Connection errors in components or cables can cause circuits to malfunction or fail.

Lack of testing: Failure to test components or circuits as you go can result in difficulties that are more difficult to diagnose later.

Safety precautions when handling electronic components:

Avoid static energy by using an anti-static wrist strap or touching a grounded object before handling sensitive components.

Switch off power: Always switch off the power to circuits before making any changes or measurements.

Use adequate tools: Use tools designed for electronics work and kccp them in good shape.

Beware of heat: Components might become hot during operation, so handle them with caution.

Troubleshooting for Electronic Circuits

Check power: Make sure the circuit is receiving power and that the power supply is working properly.

Inspect components: Look for broken or poorly placed parts.

Using a multimeter, measure voltages and resistances to find problems.

Divide and conquer: Partition the circuit to isolate the issue location.

Check connections: Confirm that all connections are secure and proper.

Resources for more learning and project ideas

Online resources: Websites such as SparkFun, Adafruit, and All About Circuits provide tutorials, projects, and discussion boards for learning and sharing.

Electronics kits, such as those from Arduino or Raspberry Pi, provide hands-on learning opportunities with projects ranging from beginner to advanced.

Creating a Simple Test Bench for Electronics Experiments

Required components: breadboard, power supply, multimeter, oscilloscope (optional), and jumper wires.

Setup: Place the components on a sturdy surface and attach the power supply to produce the voltage necessary for your research. Use a multimeter to measure voltage, current, and resistance, and an oscilloscope to view waveforms if necessary.

Function: The test bench enables you to swiftly prototype circuits, measure critical parameters, and effectively debug faults.

These advice and resources should assist you with your electronics projects and investigations

THE END

www.ingramcontent.com/pod-product-compliance
Lightning Source LLC
Chambersburg PA
CBHW050035230526
45470CB00003B/1283